PRAISE FOR *REWIRE!* AND MIKE J. WALKER

"*Rewire!* is an insightful guide that simplifies and demystifies digital ecosystems. This book will ground you in how to make th~ ~ight decisions for your business, starting with pi

—**Judson Althoff**

Executive Vice President, W

Business, Microsoft Corpor

"In a time of digital transformation and disruption, Mike has written a great book that can help everyone better understand how to design and build the right digital ecosystem and governance model for their organization."

—**Dr. Jonathan Reichental**

CEO, Human Future; Professor; Author

"*Rewire!* is a quintessence of how the right technology stack and business framework can be merged to build an apt digital ecosystem model and achieve desired business objectives. Drawing from his elaborate and successful career in the tech industry, Mike has put forth a comprehensive guide of actionable insights, smart tools, and novel ideas to help create and operate a digital ecosystem. This is a phenomenal read!"

—**Ravi Krishnaswamy**

Corporate Vice President, Microsoft Azure Global Industry Engineering

"As a CIO charged with looking over the horizon and preparing the organization for opportunities and threats, any guidance that can explain our new digital reality is invaluable. [In *Rewire!*] Mike J. Walker not only describes how digital ecosystems work, but how to make them work for you in your organization. With both detailed analysis and familiar examples, Mike is helping move past systems and platforms and shows us how to *Rewire!*"

—Doug McCollough
Chief Information Officer, City of Dublin, Ohio

"Convergence of technologies, the golden trifecta, digital ecosystems, and new business models—Mike J. Walker provides organizational leaders with a playbook of business transformation in *Rewire!*, a 'how-to' for riding the third wave of the digital economy."

—Scott J. Allen, PhD
Dr. James S. Reid Management Chair, Boler College of Business, John Carroll University

"Smart business leaders know that digital ecosystems are where value will be created and captured in the digital era. Mike is your expert guide to help you navigate the road ahead and design your future business model with clarity, confidence, and competence. With Mike's playbook in hand, forge your firm's foundation for the future and learn the mindset, skill set, and tool set to master digital ecosystems, grounded in real stories, real experiences, and real strategies. Don't get left behind."

—J. D. Meier

Director of Innovation, Microsoft Digital Advisory Services

"Mike is the authority on how emerging technologies manifest into next-generation business models. This book extends that thought leadership by helping businesses understand the impacts of platform business models and how they manifest into digital ecosystems."

—Ben Blanquera

Vice President, Covail

"The era of individual commercial organizations maximizing profits is over. As digital platforms continue to reduce economic, social, and geographical friction, how stakeholders across the value chain access information, goods, and services has rewired markets and industries, creating new ecosystem models of competition and collaboration (often at the same time). *Rewire!* breaks down the complexities of the new ecosystem paradigm and provides a concise yet practical playbook for business leaders to rethink their business strategies in the digital ecosystem era. A must-read for any business leader!"

—Howe Gu

Global Digital Transformation Strategy Lead, Microsoft Consulting Services

"*Rewire!* will challenge your thinking as it outlines the 'why' and 'how' of digital ecosystems. Ecosystems are the business models of the future to transform your business and create massive new value. If you are interested in giving it a shot, this insightful read takes you on the journey!"

—Anne Bruce
Industry Digital Strategist, Microsoft

"Our taXchain digital ecosystem was born with the idea to set up a unique information platform using blockchain-enabled digital ecosystems to securely exchange tax and customs data between companies and authorities. Fortunately, Mike came into play with his ecosystem analysis that gave us not only the insights about different existing solutions but also gave us a structured approach how to build the ecosystem. In this book, he provides the 'cookbook' for a coherent framework that helped me to break down the complexity into work packages, build sustainable cost and price models, and create a long-term road map for success."

—Dr. Olga Chatelain
IT Internal Processes and Tools, Siemens AG

"*Rewire!* provides practical and straightforward guidance to help you demystify your digital ecosystem and help you define a digital platform business model that could be a game changer in the twenty-first century. Read this book!"

—Valerie Olbrick

Digital Strategist, Microsoft Corporation

"The future of innovation will be about digital ecosystems, and *Rewire!* is the playbook business leaders will need to create sustainable new value for their organizations and the markets their organizations serve."

—Brian Peterson

Global Innovation Advisor, Microsoft

REWIRE!

MIKE J. WALKER

REWIRE!

USING THE DIGITAL ECOSYSTEM PLAYBOOK TO REINVENT YOUR BUSINESS

Published by Advantage, Charleston, South Carolina.
Member of Advantage Media Group.

ADVANTAGE is a registered trademark, and the Advantage colophon is a trademark of Advantage Media Group, Inc.

Printed in the United States of America.

10 9 8 7 6 5 4 3 2 1

ISBN: 978-1-64225-191-3
LCCN: 2020921674

Advantage Media Group is proud to be a part of the Tree Neutral® program. Tree Neutral offsets the number of trees consumed in the production and printing of this book by taking proactive steps such as planting trees in direct proportion to the number of trees used to print books. To learn more about Tree Neutral, please visit **www.treeneutral.com**.

Advantage Media Group is a publisher of business, self-improvement, and professional development books and online learning. We help entrepreneurs, business leaders, and professionals share their Stories, Passion, and Knowledge to help others Learn & Grow. Do you have a manuscript or book idea that you would like us to consider for publishing? Please visit **advantagefamily.com** or call **1.866.775.1696**.

For my wonderful wife, Angela, and our kids, Tyler, Justin, Ethan, and Jillian. Without you, I would have had this book written years ago.

Thank you for being my distraction.

CONTENTS

FOREWORD. XV

INTRODUCTION. 1

CHAPTER 1. 9
The Golden Trifecta of Digital Ecosystems

CHAPTER 2. 15
The Digital Ecosystem Framework

CHAPTER 3. 27
The Digital Ecosystem Method

CHAPTER 4. 35
Understanding the Market and How to
Ideate to Your Ideal Digital Ecosystem

CHAPTER 5. 47
Creating a Durable Digital Ecosystem Business Strategy

CHAPTER 6. 61
Designing Your Digital Ecosystem
Strategy and Operating Model

CHAPTER 7. 81
Establish Your Ecosystem Governance Model

CHAPTER 8. 97
Wrap-Up

RESOURCES 105

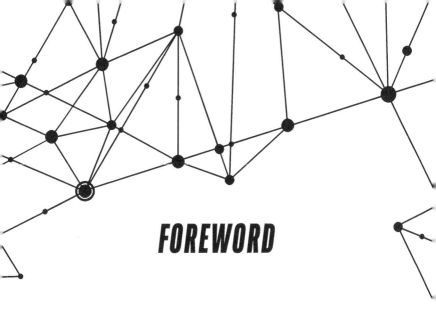

FOREWORD

From an early age, I recall doing research at the library utilizing the Dewey decimal system, the long searches for a specific book or article, all designed to speed my work along. The time it took and the latency of the material contents all resulted in a world that moved slowly and was not very "up to date." I was able to set the pace. Fast-forward to today, when information is available in real time using tools like Google. To keep up, we all must increase our speed and connections. No longer am I setting the pace, but the pace is being set for me—and it's only increasing. What changed?

Fortunately, I was engaged in the application software space from its early beginnings, and I have seen firsthand how digital technologies disrupt entire industries. With the COVID-19 pandemic, the pace of change has only increased. No longer are big tech companies such as Microsoft, Apple, Facebook, and others the only ones in the digital business. Today, regardless of the size or type of organization, whether a publicly traded for-profit entity or an institution of higher education, we are all in an era of digital transformation. The need for speed has never been greater. For those organizations that did not begin with digital technology as their cores, many are now asking how to get started.

Today's constantly changing business environment requires a highly tech-literate operational approach, regardless of the product or service. In his book *Rewire!: Using the Digital Ecosystem Playbook to Transform Your Business*, author Mike J. Walker helps you understand this environment and provides a step-by-step approach for building a digital ecosystem. His framework will help you navigate the transformation for your organization with customers, partners, adjacent industries, and even your competition with successful insight.

Whether you are at the beginning of your next-generation business model or you are seeking to take the next step, Mike does a great job of describing the steps required to form a successful digital ecosystem. His straightforward strategies provide a framework for how organizations can approach and develop their own digital ecosystems.

By simplifying and demystifying digital ecosystems, *Rewire!* is an insightful guidepost for how emerging technologies can evolve into next-generation business models. The strategies are pragmatic, and the journey will be exciting. Read this book! It will help guide you in digitally transforming your business.

—David J. Adams

Chief Innovation Officer, University of Cincinnati
www.cincyid.com

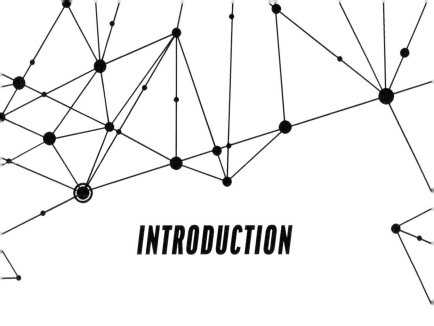

INTRODUCTION

Our philosophy is that we want to be an ecosystem. Our philosophy is to empower others to sell, empower others to service, making sure the other people are more powerful than us. With our technology, our innovation, our partners—ten million small business sellers—they can compete with Microsoft and IBM.

—JACK MA, CEO OF ALIBABA

What are digital ecosystems? A simple approach is to think of them through analogies. In general, digital ecosystems are a lot like gyms: you join one, and you have access to a variety of amenities in a shared environment. Everyone has their own personal goals, and they will use these sets of shared resources in the way that is most valuable to them. This utilitarian use of equipment and services saves everyone money and effort while creating a support system they otherwise wouldn't have. In other words, the sum is greater than the parts.

At a basic level, this is how digital ecosystems (DEs) are used. However, DEs leverage digital technologies like artificial intelligence, blockchain, and the Internet of Things (IoT) rather than physical resources. This allows participants within digital ecosystems to create value though the ability to introduce collaboration across different companies, find new or optimize existing products and services, or provide new business models to customers.

Why are we having this conversation? Today's business world flings headlong at light speed, driven by innovations in technology that change by the day. We all know that. Traditional business models are rapidly evolving as well. Are you ready?

Traditional business models are rapidly evolving as well. Are you ready?

Over the past five years, I've provided strategic advice on digital transformation initiatives to hundreds of customers around the globe, helping them innovate and maximize their business impact by leveraging digital ecosystems.

Through those initiatives I've had the good fortune to see the best and worst practices and implementation patterns, and gain a keen understanding of how to maximize value though digital ecosystems. Through over five thousand client engagements, I've published more than 150 research notes covering digital ecosystems and related topics.

Ten years ago very few companies were embracing digital ecosystems, and in many cases, existing business models didn't scale to digital—data is fragmented and vulnerable, information is inconsistent, interactions are not well supported, and one intermediary essentially controls everything. But change is underway. Today, the majority of the planet's top ten companies are digital-ecosystem-based—Microsoft, Apple, Google, Amazon, Alphabet, and Alibaba, along with rising stars such as Uber and Airbnb. Digital ecosystems enable those companies to do what they do so well, and while they might not yet be household or boardroom names, digital and platform ecosystems are the predominant business models driving the future of commerce.

Andy Rowsell-Jones, vice president of Gartner Research's CIO and executive leadership research team, emphasizes the importance of DEs in today's companies: "Digitalization in most enterprises is maturing, and as it does, it becomes more likely that these enterprises will become part of a digital ecosystem." Digital ecosystems, he says, "enable you to interact with customers, partners, adjacent industries, and even your competition."[1]

If you've never heard of digital ecosystems, you may be thinking this is the best-kept secret in the business world, and you'd be partially right. A recent Gartner study revealed that 79 percent of the top-performing modern organizations participate in a digital ecosystem, while fewer than half of the "average performer" companies have made the leap.[2]

Technical advances present endless connections through which everything is talking, brought directly to our office desks and into our hands by the mobile, wearable sensor technology we use and rely on every day. The ubiquity of these connections is enabled by a wave of communications technologies advancing deliberately over the past five to ten years around

---

1   Kasey Panetta, "The Gartner 2017 CIO Agenda urges all organizations to seize the digital ecosystem opportunity," Gartner.com, October 17, 2016, https://www.gartner.com/smarterwithgartner/ecosystems-drive-digital-growth/.

2   Madeline Bennett, "What is a digital ecosystem, and how can your business benefit from one?" *The Telegraph*, April 12, 2017, https://www.telegraph.co.uk/business/ready-and-enabled/what-is-a-digital-ecosystem/.

5G technologies, the latest Bluetooth advances, and Wi-Fi capabilities that make it possible to do very sophisticated things. The interoperability between different technologies, companies, and platforms creates an expectation of complete connection from social, people, and customer perspectives.

We see this happening in our day-to-day worlds, and we are accustomed to it, but the "how" is less understood. The key player on this global business stage is digital ecosystems. Formally defined, digital ecosystems (DEs) embrace the complexity of endless connections in self-organizing, dynamic, adaptive ways that leverage advanced technologies to maximize or create new value-exchange networks.

Consider the big names like the aforementioned Google, Apple, and Amazon, all wildly successful examples of digital ecosystems created to meet consumer demand. It's that insatiable demand that's forcing separate entities to collaborate. As Gartner puts it, digital enterprises are essentially "an interdependent group of enterprises, people and/or things that share standardized digital platforms for a mutually beneficial purpose, such as commercial gain, innovation or common interest."[3] This doesn't mean they

---

3   Panetta, "The Gartner 2017 CIO Agenda urges all organizations to seize the digital ecosystem opportunity."

become one ecosystem to rule them all; we live in a capitalistic society that incentivizes business models that provide unique value to customers, and those models must interoperate with local government, city infrastructure, competitors, and myriad other influencers. And in this increasingly connected world, we generate more data than ever before, and the risk of data breach has significantly increased. This is where the qualities of a well-designed digital ecosystem play a huge part. Leveraging technologies like blockchain and artificial intelligence (AI) enables digital ecosystems to mitigate or outright eliminate risks by introducing sovereignty of data, data self-governance, data anatomization, algorithmic encryption, data immutability, and transparency.

> In this increasingly connected world, we generate more data than ever before, and the risk of data breach has significantly increased. This is where the qualities of a well-designed digital ecosystem play a huge part.

However, this piece of the puzzle is built around a massive convergence of technologies that are built on each other, rather than used in isolation. The golden trifecta of these technologies, introduced in the first chapter, is 1) artificial intelligence—the brains behind data, 2) the Internet of Things—bringing information in the physical world to a digitized format, and 3) blockchain—a unifier of data that is essentially an element of trust holding all of this together. This technology trio helps us reach a point of creating viable digital ecosystems and platform business models, similar terms that describe a convergence of technologies.

Here's an example: in recent years, blockchain was an emerging technology with enormous application potential, but it was hardly mainstream. Through a digital ecosystem lens, we learn that blockchain is a tool that enables specific capabilities from pharmaceuticals to aviation to industrial manufacturing and, of course, everyday online commerce giants. It's possible to build a digital ecosystem without blockchain, but it is a critical component that weaves a connected and viable business framework.

Today's business environment requires a highly tech-literate operational approach, regardless of product or service. Companies targeting a successful future will no longer be able to "work around" this reality, which is already rapidly taking shape.

In this book we will cover two major aspects of digital ecosystems. First, the Digital Ecosystem Framework defines the "what"—a customizable reference model of all the components that comprise a digital ecosystem. Having a reference model only shows us the pieces and parts but doesn't tell us how to build. Second, a repeatable and predictable methodology provides the fit-for-purpose guidance on how to build your own digital ecosystem; it's a separation of what it is, the parts, and how to build it. Consider the difference between *Gray's Anatomy* (the original medical textbook, not the TV show) and WebMD. The former tells us what all the pieces are, and the latter understands what to do with them.

Simply stated, this is a business transformation book, introducing tools to help demystify digital ecosystems with real-world examples of companies using the technology today and actionable recommendations for future-proofing your company. The book focuses not on the technical nitty-gritty, but instead examines strategy and formation, and as shown in the graphics to follow, readers should expect to be able to fully understand how to strategize and form a digital ecosystem. Think of this as a digital ecosystem primer, with accessible links to additional resources.

Let's get started.

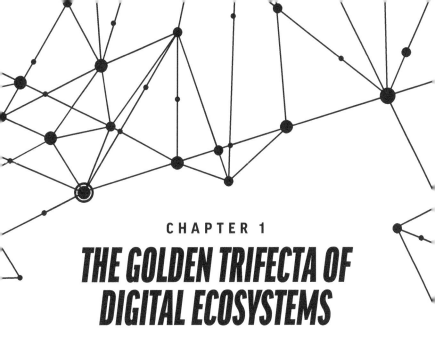

CHAPTER 1

THE GOLDEN TRIFECTA OF DIGITAL ECOSYSTEMS

E cosystems are not a new paradigm. They were first referenced in a 1993 *Harvard Business Review* article by James Moore titled "Predators and Prey: A New Ecology of Competition." Moore discussed the concept that companies must innovate with technology and evolve capabilities around emerging business interaction patterns. But in 1993, we had just begun the internet revolution.

Today, the digital revolution has ushered in a number of innovations that have been game changing for digital ecosystems, and it's not just one technology on its own. The convergence of ubiquitous sensors, availability of communications, trust-based information exchange, advanced intelligence, and virtually limitless cloud computing has been a driving force for digital ecosystems. While ecosystems have been in existence for some time, the use of digital technologies takes ecosystems to a whole new level.

The convergence of ubiquitous sensors, availability of communications, trust-based information exchange, advanced intelligence, and virtually limitless cloud computing has been a driving force for digital ecosystems.

There are three technologies fueling this convergence that I like to call the Golden Trifecta of Digital Ecosystems. This includes the IoT, AI, and blockchain. One way to think about this trifecta is to picture these technologies like the human body. IoT uses all types of sensors to capture data from the physical world, much like our own nervous system, translating the world though our fingers, eyes, and ears. Blockchain is akin to our circulatory system; it provides a protocol for the integration of information like our veins, a trust-based, immutable ledger as a storage mechanism similar to our DNA, and it's algorithmically protected by encryption, much like our immune system protecting from foreign invaders. Lastly, all roads lead to AI, which provides intelligence to trusted sensory information, much like the human brain.

IoT provides vital data into the ecosystem in real time without any human intervention. Interestingly enough, Gartner Research found that 75 percent of organizations that are implementing IoT technologies have already implemented blockchain or will implement by the end of 2020.[4] Using blockchain in conjunction with IoT provides compelling new ecosystem capabilities that enable interactions between people and things, which may render existing models ineffective or obsolete.

Enabling a digital ecosystem with blockchain creates a collaboration-based value network between organizations that harnesses unique capabilities of blockchain, such as data self-sovereignty, tamper-resistant data, peer-to-peer data collaboration, and distributed governance.

AI brings a vital capability to ecosystems: intelligence. With new contextual data from IoT and trustworthy, cleansed data from blockchain, AI can be particularly useful. Using AI as a shared digital ecosystem

---

4   "Gartner Survey Reveals Blockchain Adoption Combined With IoT Adoption Is Booming in the U.S.," Gartner.com, December 12, 2019, https://www.gartner.com/en/newsroom/press-releases/2019-12-12-gartner-survey-reveals-blockchain-adoption-combined-with-iot-adoption-is-booming-in-the-us.

capability across a diverse set of actors and data is truly transformational. In fact, many companies have already started. Deloitte Global predicts that among companies that adopt AI technology, 65 percent will create AI applications using cloud-based development services.[5]

The shift from isolated technical infrastructure to platforms enabled by digital ecosystems lays the foundations for entirely new business models. Arguably, any one of these technologies can provide value on their own. However, the real value is when these three converge to create an ecosystem solution. This allows organizations to:

- capture unprecedented amounts of data that can be enabled by predictive models to create new revenue streams and market differentiating products;

- establish new business-integration patterns to create compelling new user experiences;

- establish a product or service with minimal operating costs or assets while entering new markets virtually friction-free, similar to Airbnb or Uber's ecosystem capabilities; and

---

5   "Tech Trends 2019," Deloitte Insights, 2019: 23, https://www2. deloitte.com/content/dam/Deloitte/fi/Documents/technology/ DI_TechTrends2019.pdf.

- harness the power of the capabilities of mobility, much like how Apple has democratized application development with the Apple App Store.

With a background on their technical side, we are ready to dive into the "personality" and framework of digital ecosystems. What are they, and how do they work?

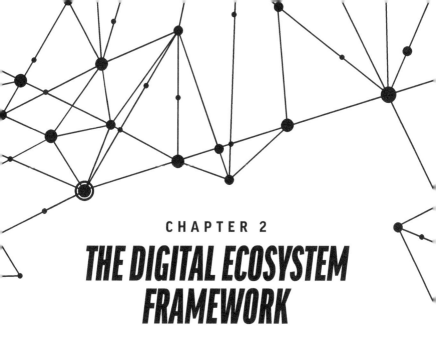

THE DIGITAL ECOSYSTEM FRAMEWORK

C onsumers see digital ecosystems in action every day, but it's like the songwriter behind the music—something is working hard to make the entire scenario come to life. Purchase an item on Amazon, and a complex array of digital forces go to work from the shopping stage to your doorstep. Most people have no idea what is involved, only that it works. Even savvy business owners may only have a fringe understanding of the concept.

Through my research I've identified over ninety digital ecosystems in various forms of development. These ecosystems can vary in size, type, industry, location, and business focus. For example, there are global ecosystems like those already mentioned— Apple, Uber, Airbnb. There are also industry-specific ecosystems including B3i, RiskBlock Alliance, Hyperledger Healthcare Working Group, Climate Change Coalition, Mobility Open Blockchain Initiative (MOBI), taXchain, and HealthPass.

One way to grasp the concept of DE frameworks is reflecting on natural ecosystems. A complex yet symbiotic network of flora and fauna work seamlessly to keep the earth ticking along. Digital ecosystems work the same way; complex networks enable value exchanges between new and existing business partners. Think of it as an intricate collection of companies coming together in a unified approach to a particular market. To that end, it is important to have a framework to understand how to design, operate, govern, and derive value monetarily or through cost optimization. We will examine this

> **A complex yet symbiotic network of flora and fauna work seamlessly to keep the earth ticking along. Digital ecosystems work the same way; complex networks enable value exchanges between new and existing business partners.**

closely throughout the book.

For a group of organizations driving toward a unified set of goals, you need a cohesive framework made of six vital aspects defining what is required to successfully deliver a digital ecosystem—governing organization, capabilities, participants, governance rules, value realization, and ecosystem operations. Keep in mind that each of these aspects is required and subsequently dependent on the others. As shown in the following graphic, meeting desired goals relies on a governing organization housing an ecosystem's key decision authority. This is where vision, strategy, and business model establishment happens, manifesting in different ways, depending on driving parameters. We see these parameters as the four middle wedges in the graphic—capabilities, value realization, participants, and governance rules.

Figure 2.1

Governing Organization: The decision maker and business authority that defines the ecosystem strategy, governance, go-to-market, and policy, and provides oversight into the vision, economics, and capability development of the ecosystem

Capabilities: Defines the business capabilities, their scope, delivery, and the introduction timing within the ecosystem

Participants: Identifies the participant membership criteria, acceptable behaviors, and activity governance

Governance Rules: Establishes the governance framework that will guide all decisions made within the ecosystem based on the benefits and risks they pose

Value Realization: Quantifies and defines the value streams for all participants, risk tolerance levels, and digital ecosystem success criteria

Ecosystem Operations: Defines the operational model, the technologies and assets delivered, required competencies, and the specific delivery organization to operate

---

Based on what you ultimately want to deliver, capabilities define the key features and functions differentiating an ecosystem, as well as identify derived value in terms of financial value or risk mitigation. Value realization looks at transformational aspects or new business models, and potential optimization within existing. This all goes on to inform the types of participants, which are generally tied directly to their capabilities' interest areas. There are many types of participants, including both people and virtual roles, and each will have different value realizations.

As with most everything, governance must be in place to ensure that participants act in the principle and spirit of the ecosystem to mitigate the risk of "bad actors" in the system and establish trust.

In the middle is the ecosystem operations component stitching it all together. This is day-to-day management of services, onboarding new participants, solution development within the ecosystem, and any required change management.

The importance of a Digital Ecosystem Framework cannot be stressed enough. A robust and detailed framework provides you with a solid foundation of understanding around key digital ecosystem concepts. Concepts like vision, business models, value models, ecosystem models, risk models, information architecture, and governance models are a few of the vital concepts that must be understood.

> **The importance of a Digital Ecosystem Framework cannot be stressed enough. A robust and detailed framework provides you with a solid foundation of understanding around key digital ecosystem concepts.**

It is critical to also understand that a Digital Ecosystem Framework identifies key aspects to be addressed and that the ecosystem can be customized to its accompanying method. That's where the real power lies—putting DE frameworks (DEFs) into action by leveraging the Digital Ecosystem Method (DEM), which we'll discuss more in the next chapter, to identify 1) new strategies, the "why," 2) a scalable business model, the "what," and 3) an optimal operating model, the "how," with a potential set of companies.

Every DE should look different—even with similar capabilities. For example, some digital ecosystems with the same capabilities work best in a completely closed process to preserve their competitive advantage, while others choose a semi-open model. The framework provides the core questions you must ask to establish a fit-for-purpose ecosystem.

Consider the real-world example of the logistics industry. This highly integrated and pervasive capability is implemented across most industries, covering everything from a resource's origin to a finished good's destination. Globally, the logistics industry represents approximately 12 percent of the entire world's GDP![6] As technology advances, so do incumbents like Amazon who are willing to embrace these technologies and new business approaches.

The figure below shows how Amazon and others are building digital ecosystems, exposing a brittle and linear supply chain that is rampant with process inefficiencies, fraud, and the inability to quickly adapt to customer demands.

---

6   Todd Maiden, "How big is the logistics industry?" Freightwaves.com, January 11, 2020, https://www.freightwaves.com/news/how-big-is-the-logistics-industry.

Traditional Value Chains
Linear and Compartmentalized

Digital Ecosystem-Enabled
Orchestrated and Disintermediated

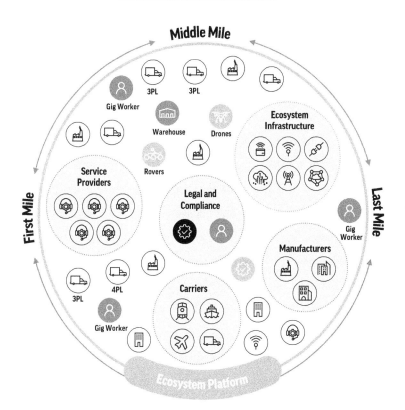

Figure 2.2

Take the food and drug global market, where roughly $200 billion of counterfeit drugs flow through the logistics supply chain.[7] Federal governments are now getting involved and requiring ecosystem-level visibility of goods to ensure authenticity. This is a big deal—digital ecosystems provide the capability to prevent inefficiencies and fraud and unlock significant cost savings that can be passed on to the consumer.

Within the DEF, governance rules are defined for each workflow, such as temperature parameters, digital signatures, changes in management board, and similar aspects. The ecosystem operations core is managed by the governing entity and in this case will evolve to a third party in the near future.

With the right technology, creating a DE framework like this should be really easy. However, defining and reaching agreement on a DE framework is difficult, demands patience, and requires diverse roles across many companies to make very tough decisions impacting entire industries. Often, rather than technologists, these roles are decision makers such as lawyers and executives.

With a digital ecosystem skeletal structure and

---

7   "How Blockchain Will Accelerate Business Performance and Power the Smart Economy," *Harvard Business Review*, October 27, 2017, https://hbr.org/sponsored/2017/10/how-blockchain-will-accelerate-business-performance-and-power-the-smart-economy.

framework to define the "what," we need an accompanying Digital Ecosystem Method to help us with the "how" to tailor to your needs.

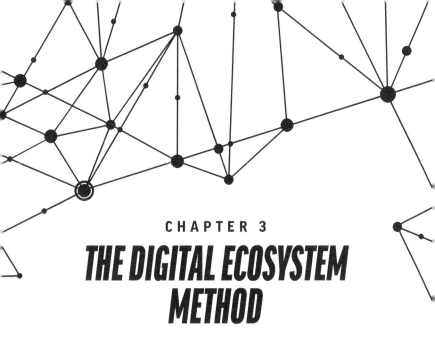

CHAPTER 3

THE DIGITAL ECOSYSTEM METHOD

My team and I embarked on creating an exchange for securely managing tax and customs of goods between companies and authorities called taXchain. I can tell you that trying to do this on your own without a structured approach is a recipe for disaster. Using the Digital Ecosystem Method provided the "cookbook" to rapidly accelerate our ecosystem creation while avoiding some major hurdles!

—DR. OLGA CHATELAIN, SIEMENS AG

n the world of digital ecosystems, there are three major phases of the journey—first, defining the strategy; second, incubating and implementing the production-grade solution; and finally, maintaining and governing it. This chapter dives into the end-to-end method that will provide key definitions and overall context into using the Digital Ecosystem Method.

The Digital Ecosystem Method is a customizable digital strategy and transformation approach that enables business and technology leaders to create innovative digital ecosystem business models to new or existing markets. This highly iterative method provides an actionable digital ecosystem strategy, business model, operating model, and transformation road map.

Let's start with by looking at the first stage of DEM—envision. This is an incredibly important part of the process, because "you don't know what you don't know." This part of the process is about taking a step back and challenging your strategy with new inputs taken from market trends, new technology innovations, and compelling new experiences leveraging them both.

During the envision stage, you're taking the time to make sure uncertainty doesn't steer you down a path where unnecessary risks are introduced, or you miss out on the next big opportunity. Trendspotting allows us to fully understand market trends and the potential of digital ecosystems, not only within today's markets but also tomorrow's. Once those trends are understood, qualifying them through ideation allows you to envision the art of the possible.

As we move to the strategize stage, we've eliminated quite a few bad ideas, technologies, and market approaches. This is an important input, as it provides a well-informed and durable strategy with strong due diligence and a solid understanding of what business and technology advancements can be utilized when evaluating whether a strategy needs revision or whether a new strategy is needed. The focus of this stage is to clearly define the ecosystem vision, value proposition, formation requirements, business governance, economic models, and the operating model that future-proofs your ecosystem business model.

At the architect stage, major technology choices are made, and a road map is designed based on the first two stages, resulting in a company-specific business model. These technology choices provide the final set of information to finalize the governance model. While the business governance and ecosystem governance are nearly complete, outstanding questions in technical governance and other governance are answered.

Strategy & Formation

Envision

Ideate, envision, and identify digital ecosystem opportunities through open innovation approaches.

1.1 Understand both the internal needs but also the external trends in the market and the unique capabilities digital ecosystems provide.

1.2. Continuous developments of MVE pilots to prove business and technical viability.

1.3. Ideate and envision the art of the possible to identify new and unknown opportunities or refine an existing business process to maximize value.

Strategize

Strategize defines the value stream that will future-proof the ecosystem business model and all major decision points.

2.1 Rationalize your existing business and envision the digital business strategy and ambitions.

2.2 Model the ecosystem business strategy and create the business strategy along with all the economic factors for participants.

2.3 Assess inhibitors and risks to participants and to adoption.

Architect

Tuned operational model and governance with clearly defined business, technology, and human processes in place to maximize value and growth.

3.1 Establish the ecosystem operating model and ecosystem governance rules.

3.2 Architect ecosystem solution architecture based on the ecosystem design.

3.3 Plan the transformation road map for controlled capability rollouts to participants.

3.4 Finalize the business, ecosystem, and technical governance models.

MVE Approach

Iterative architecture and development of the strategic road map that develops rapid MVE prototypes and controlled production roll-outs.

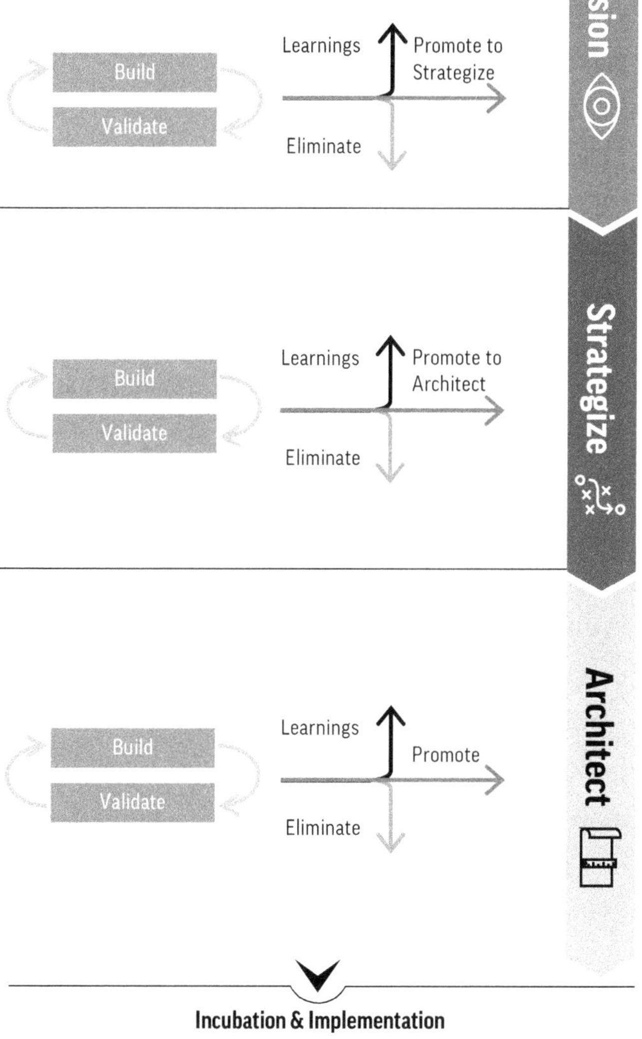

Build
Validate

Learnings
Promote to Strategize
Eliminate

Envision

Build
Validate

Learnings
Promote to Architect
Eliminate

Strategize

Build
Validate

Learnings
Promote
Eliminate

Architect

Incubation & Implementation

Figure 3.1

31

Keep in mind the tendency to address, and the inherent challenge of addressing, future topics too soon (design before strategy), which almost always results in losing sight of the ecosystem's true nature. There hasn't been an opportunity to mull things over with stakeholders, and it's not just your company in the picture; multiple companies must be in agreement. Tight alignment with ecosystem participants is vital to success, and in many cases, this involves companies with little to no knowledge of digital ecosystems.

> **Tight alignment with ecosystem participants is vital to success, and in many cases, this involves companies with little to no knowledge of digital ecosystems.**

Siemens and Henkel, two of the world's largest manufacturing and consumer goods companies, for example, had arrived at a new chapter of their digital journey. They knew digital ecosystems were the answer but needed help building a strategy maximizing the power of DE capabilities with their own goals and objectives. Their collective industry is highly regulated but still largely run with paper-based transaction reconciliation and verification, in addition to being wildly human-intensive and fraught with potential error. They needed a method they could trust to address business, legal, and financial issues head on.

They realized that digitalization is the fourth industrial revolution, where digital and physical worlds merge in tandem with their respective economies. In that arena, their teams developed ambitious goals to create an inclusive digital ecosystem called taXchain, using DEM to reinvent and digitize tax and customs handling of global goods across global supply chains. This methodology brought together two companies that had not previously worked at this level of detail or code development and leveraged blockchain technology to create a durable digital ecosystem strategy. From a trade finance perspective, this allows them to understand what goods are shipped across borders and how those goods should or shouldn't be taxed. To date, this has been a labor-intensive thorn in the industry laced with paper-based cost overruns and tracking issues. DEM digitizes everything and keeps industry participants compliant with inherent regulations. It's an open ecosystem welcoming all participants to ideate on a wider scale of possibility.

With a solid grasp of a digital ecosystem's method, you can look closely at your particular market to gain full understanding to prepare an effective approach to ideation.

CHAPTER 4

UNDERSTANDING THE MARKET AND HOW TO IDEATE TO YOUR IDEAL DIGITAL ECOSYSTEM

As industries shift from monolithic, centralized business models toward more dynamic and open ecosystems, the approaches required for success change. If you are going to survive and thrive in this environment, you must have an innovation mindset and a technology trendspotting capability that ensures you are staying ahead of the curve. I believe that understanding emerging technologies now is critical to being able to empathize with your future customers and the world they will live in tomorrow. This kind of empathy has the potential to unlock innovative ideas for new products all across a company, and working with Mike J. Walker to mature our trendspotting approaches has helped us complement our existing innovation practices in a material way.

—MIKE FULTON, NATIONWIDE INSURANCE

W e've now entered the third wave of the digital economy that has reinvented ecosystem-based business models. The first wave was based primarily on the internet and e-commerce. The second was enabled cloud computing, mobility, and big data. This third wave blends the physical and the virtual by building on the previous waves by adding intelligence to data with AI, making endless connections of people and things with 5G, securely managing data with blockchain, and equipping a wide variety of devices and robotics with IoT.

Historically, each of these waves of innovation has challenged the business model status quo. The problem for many businesses is that they just can't go digital overnight. This is due to addressing large and looming questions around the company's digital strategy, determining the right mix of new capabilities versus improving existing ones, and analyzing the company's fundamental vision.

Digital ecosystems fundamentally disrupt traditional business models, and while you may have the impression that you have most of the answers, or perhaps feel like you know the "right thing to do," you most likely don't. The fact is that decisions must come from many people across a variety of roles. When we think about what our digital ecosystem should look like, we must take an inclusive, innovation-based approach, given the wide range of business and legal questions to be answered before the technical. The goal throughout is moving from the most abstract thought processes, the craziest ideas, to something viable for your company.

The first stage to DEM that we talked about in the previous chapter—envision—focuses on bringing inclusive innovation

> When we think about what our digital ecosystem should look like, we must take an inclusive, innovation-based approach, given the wide range of business and legal questions to be answered before the technical.

instead of a narrow and dogmatic approach, focusing on bringing in a diverse set of ideas from many different participants from within and outside your organization, as shown in the following graphic.

We take an open, innovation-based approach to look at broad market trends and activities with a systematic way of distilling those into key opportunities for your digital ecosystem.

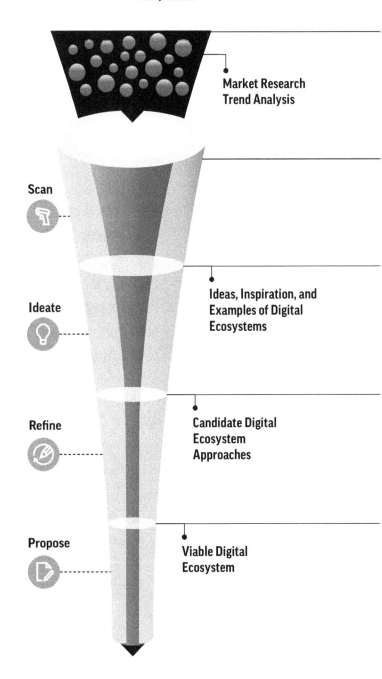

Market Research
Trend Analysis

Scan

Ideate

Ideas, Inspiration, and
Examples of Digital
Ecosystems

Refine

Candidate Digital
Ecosystem
Approaches

Propose

Viable Digital
Ecosystem

TRENDSPOTTING

- Market dynamics
- Business, technology, behavioral, societal, economic, health, legal, and geopolitical trends
- Emerging technologies
- Existing solutions
- Research

IDEATION

Scan

- Identify existing digital ecosystems
- Analyze DE market drivers
- Understand DE business models
- Determine primary technologies to focus on

Ideate

- Idea identification
- Idea categorization
- Idea prioritization
- Opportunity viability

Refine

- Continuous iteration and refinement of ideas
- Validate patterns; organize and group related ideas
- Consolidate redundant ideas

Propose

- Idea Viability Worksheets
- MVE opportunities will connect how this benefits the customer experience, impacts the value of the chain / creation model, and assess the alignment to current strategic goals and objectives

Figure 4.1

In this DEM stage, efforts are concentrated in a pragmatic way to include business decision makers—your vital constituency. These are essentially the people who write the checks (or go to jail when bad things happen). What used to be "build, design, and present to stakeholders" is now adapted to embed all players early in the process so they understand market dynamics, technologies, and trends, and are part of decisions in real time. Another added benefit is establishing a level of shared accountability that embeds those decision makers in the creativity process.

You might think this could get unwieldy very quickly, and you're right. Therefore, DEM prescribes that the approach is taken pragmatically by starting with a deliberate trendspotting activity followed by piloting multiple minimum viable products (MVPs) and conducting structured ideation workshops. This allows your team to complete due diligence on all market options and conditions that provides them with a well-understood marketplace of disruptions and trends into the ideation process. MVPs allow you to validate opportunities in a risk-averse way, quickly and cost effectively.

In the context of digital ecosystems, your innovation team has ideally already done some trendspotting, but if that's not the case, this work falls on to your team's to-do list. If your digital ecosystem requires a detailed trendspotting stage, this is a good thing. It arms all participants with the same type of marketplace knowledge. I've found that this significantly reduces confusion from participants with various degrees of understanding of the market and the enabling technologies and terminology of those areas.

Trendspotting in its entirety is quite comprehensive, so for the purposes of this book, we'll focus on macro-approaches and extended resources. Expect to conduct the following three stages in your trendspotting efforts:

1. **Scan:** Conduct market research to gather information from various sources, including industry analysts' perspectives on DE, trade associations' standards, existing digital ecosystems and use studies, along with educational material from vendors or academia about enabling technologies like AI, IoT, and blockchain.

2. **Select and analyze:** From the thousands of technologies, trends, and innovations, synthesize them down to a set of trends and technologies that most apply to your digital ecosystem needs.

3. **Recommend:** With further business qualification and viability assessments, provide a set of

recommended trends and technologies that can be used as an input into your digital ecosystem ideation efforts.

Sometimes trendspotting can take two days or two weeks; it all depends on the level of knowledge your team has around market trends.

Starting with a trendspotting stage prior to ideation provides your team with an outside-in perspective that can fuel meaningful ideation sessions.

Starting with a trendspotting stage prior to ideation provides your team with an outside-in perspective that can fuel meaningful ideation sessions. Every digital ecosystem ideation session looks a little different based on several factors that will guide your ideation approach. These can include:

- Breadth of impact on your business: Do you want to create something highly transformative to your company and market, or will you optimize an established process?

- How well you understand the problem you're trying to solve: Your approach will be based on that understanding.

- Your organization's appetite for change and degree of innovation culture: This will affect how you approach people involved in the innovation, what they're willing and unwilling to do, and types of activities they will face.

- Your company's level of focus: Proper focus helps provide adequate scope for your innovation efforts and informs applicable activities.

Taking this approach avoids the classic trap of letting methodology dictate you instead of you dictating the process. This brings people together with a big catalog of ideas from the marketplace to decide what to do with them. These inputs will influence four major steps to any digital ecosystem ideation:

First, use your trendspotting results to understand the ecosystem potential. If a scan hasn't been done, understanding how emerging technologies are used to create digital ecosystems will show your company's level of impact.

Second, ideate on the possible. Typically conducted in a workshop setting, these workshops bring together key business and technology leaders to brainstorm on where they can optimize or transform their business. Expect the majority of work to happen at this stage. Activities that are designed to determine the digital ecosystem viability will be done through idea identification, categorization, and prioritization.

What are some trade-offs and activities involved when it's time to refine your ideas? Do you join an existing system and innovate within or create your own?

Third, to refine ideas to maximize impact and to fully qualify an idea's business opportunity, it is critical to continuously challenge and remove waste from the ideas generated.

Finally, propose the minimum viable ecosystem. The ability to shift as quickly as possible into actionable opportunities is vital in this fast-paced digital world. However, we want to ensure that ideas generated are actual business opportunities with a focus on strategic relevance and overall value potential. Using tools like Idea Viability Worksheets (you can view a sample of this in the resource section at end of the book) puts these technical solutions into a business context that senior business and technology leaders can understand and support.

The ability to shift as quickly as possible into actionable opportunities is vital in this fast-paced digital world.

This approach stitches everything together so you have a full understanding to help us understand market drivers, key trends and players, and available technologies. You've worked with business partners and subject matter experts to ideate around the state of possibilities and included them all to verify what is viable for the company in terms of risk levels. Ultimately, these contextualized ideas become digital ecosystem opportunities that in turn become capabilities presented to market.

The overarching point is to quickly figure out what's not going to work. In fact, failing fast is very important in successfully executing methodology because people naturally gravitate toward things they can touch and feel, and it's easier making decisions that way than looking at topics on a PowerPoint slideshow. Our goal is to understand market focus and key trends, identify available technologies, and refine ideas pursuant to high value potential.

Before we make any material investments in trying to either optimize or reinvent our strategy, we need to make sure that we understand all the opportunities that a digital ecosystem can present to our organization. That will include us making sure that we look at the market, the conditions of that market, the technologies available to us, and our relationships with our partners and where they're going. When we understand all of these conditions that are outside of our control, we can ensure we have the information we need to make informed strategic decisions going forward.

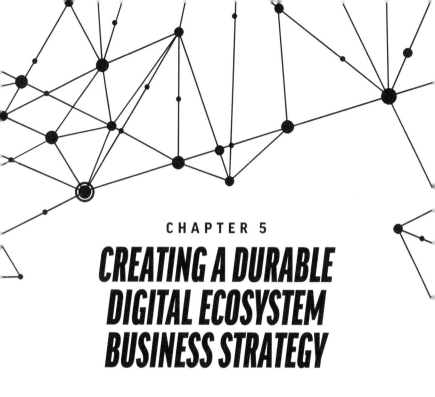

CHAPTER 5

CREATING A DURABLE DIGITAL ECOSYSTEM BUSINESS STRATEGY

N ow that we understand what's possible and we've done our due diligence on the market's viability, we need to start with the purpose, the "why" of creating this ecosystem to provide us with a road map of questions to ask going forward. This helps avoid unnecessary assumptions that may introduce risk or potential failure into the implementation of an ecosystem.

Digital ecosystems are not only disrupting companies' current ways of doing business, but they also force them to rethink many aspects of their business models. Companies are seeing how DEs can define entirely new market segments, reinvent products or services, breathe new life into existing value streams, and create enhanced experiences with customers.

So how do you get started? As shown in figure 5.1, it all begins with taking those great ideas and analysis from the envision stage to inform a new strategy or refine an existing one. Having those baseline strategic questions answered arms you with the right information to make the trade-offs needed to select an ecosystem business model. In this chapter, we will concretely define your Digital Ecosystem Framework (DEF) and its corresponding strategic building blocks.

Determining which digital ecosystem business model is right for your organization is the first big decision to make.

Determining which digital ecosystem business model is right for your organization is the first big decision to make. There are three main models you can work within: founder-led, partnership-led, and cooperative-led.

1. **Founder-led:** An ecosystem that one company controls through an exclusive set of participants. NASDAQ, for example, must be highly secured, monitored, and compliant with governmental laws. The Apple App Store is another example with many participants in the ecosystem, but Apple is the one central authority approving all apps.

2. **Partnership-led:** An ecosystem controlled by an exclusive group of companies that share decision-making authority as a joint venture. Think of the airlines' alliances; Delta is part of the Skyteam Alliance, and members like Virgin Atlantic and KLM can unite their own ecosystems with Delta. If you are on a Delta flight to London Heathrow and need an airline lounge, Virgin Atlantic will support you.

3. **Cooperative-led:** An open ecosystem that is self-sustaining and governed by its own membership. A great example here is the Risk Stream Alliance, a nonprofit entity that is open to all insurance carriers that want to collaborate using open standards. Control is in the hands of an elected board of directors, not any one given company.

Envisioning Insights and Digital Ecosystem Strategic Drivers

Envisioning Ideated Opportunities	Digital Ambitions	Ecosystem Business Drivers	Ecosystem Principles	Ecosystem MVE

Drives Digital Ecosystem Business Model Choice

Cooperative-Led Digital Ecosystem

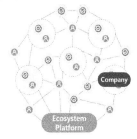

An ecosystem that is self-sustaining and governed by its members

Partnership-Led Digital Ecosystem

An ecosystem controlled by an exclusive group of companies that share decision-making authority as a joint venture

Founder-Led Digital Ecosystem

An ecosystem that one company controls through an exclusive set of participants

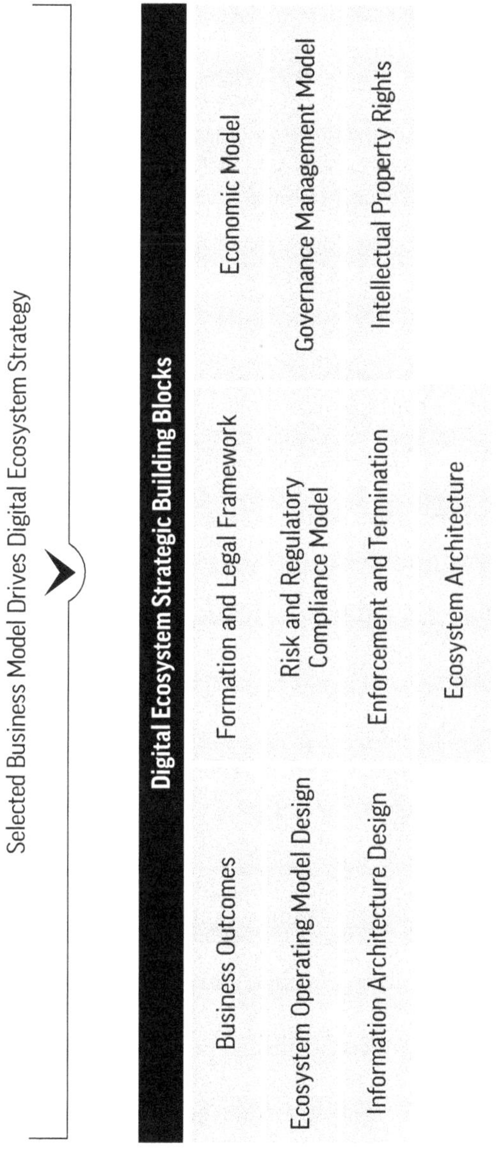

Selected Business Model Drives Digital Ecosystem Strategy

Digital Ecosystem Strategic Building Blocks

Business Outcomes

Formation and Legal Framework

Economic Model

Ecosystem Operating Model Design

Risk and Regulatory Compliance Model

Governance Management Model

Information Architecture Design

Enforcement and Termination

Intellectual Property Rights

Ecosystem Architecture

Figure 5.1
DEM business models are derived from envisioning insights
that prescribe your DE strategic building blocks.

With each of these models, there is no clearly "better" option. All three have their advantages and disadvantages based on five common considerations:

1. **Level of risk appetite:** Tolerance level to transformative business models, aggressive market position, and rapid time to market are some of the considerations.

2. **Market perceptions:** How customers, partners, and competitors react to your ecosystem will determine the business model you choose.

3. **Adoption considerations:** Identify questions of how hard it will be for participants to adopt and adapt based on the level of reinvention of their business processes and integrations.

4. **Time to value needs:** Stakeholders will need to decide how aggressive they want to be in the time in which it will take to generate value from the ecosystem.

5. **Competitive strategy:** How competitive you desire to be is one of the most influential characteristics on how open or closed the ecosystem business model will be.

Let's look at how these considerations influence the decision of each business model choice.

Founder-led business models should be considered when one organization needs full control to reinvent an existing or new business model, or control the direction of an industry or speed to market. However, avoid these if there are concerns of a single point of failure, talent acquisition, long-term founder business executive sponsorship, and funding.

Partnership-led business models should be considered with two or more companies that are not competitive or that can reasonably cooperate with compatible market motivations, partnerships, and value exchange. This model spreads the risk and monetary structure, but steer clear of these if you can't find dependable partners or if the complexities over co-investment, joint development, and intellectual property become an issue.

Cooperative-led business models should be considered when it is desirable to join a self-sustaining open ecosystem or collaborative that supports cross-company integration standards or defined information exchanges, and are platforms to unite an industry around a common set of capabilities. This is a viable option for those who don't want to build their own ecosystem but can join another to avoid start-up costs, support staff, solution design, and recruitment of members. On the other hand, choose otherwise if there are concerns over speed to market, complexities of multiple participants, or slower innovation due to getting consensus from the membership.

With a business model in hand suiting your company's needs, that decision drives all other choices going forward, but don't be surprised to run into a chicken-versus-egg scenario, with some natural back-and-forth between defining your DE strategic building blocks and your DE business model. If you're stymied by the question of "Which comes first, the business model or the building blocks?" the answer is the business model. The model provides a foundational hypothesis in which to frame the strategic drivers, and with those firmly entrenched, you can dive into the building blocks, the essence of your model's strategic digital ecosystem plan.

> If you're stymied by the question of "Which comes first, the business model or the building blocks?" the answer is the business model.

Five particular building blocks are of specific interest in regard to designing an ecosystem strategy:

1. Business outcomes

2. Formation and legal framework

3. The Economic Model

4.   Intellectual property and Data Rights Model

5.   Enforcement and termination

First, we need to define the strategic *business outcomes*—qualifiable and quantifiable goals and objectives of a digital ecosystem with a robust vision and mission statement that communicates what we want to accomplish. Why do we have this ecosystem? What's our philosophy, and what are we trying to do? This stage is meant to inspire participation.

Defining the *formation* and *legal framework* determines all specific legal and regulatory establishment concerns. This is often overlooked in the early stages and results in rework and frustration of your stakeholders. A founder-led DE, for example, will likely mirror your company—whoever creates the ecosystem owns it, and it's another product or service offered to the marketplace. Partner-led DEs are joint venture agreements between companies with shared legal responsibilities and accountability. Cooperative-led DEs allow for a variety of options: you can establish as a nonprofit or for-profit organization, which then inspires questions about where you should base your company's home—which state or country suits the company best, and what laws apply?

The Economic Model defines the participant

value-exchange benefits, costs of ongoing maintenance, and building of additional capabilities. The DE
requires investment to establish, recruit, and build,
and a lot of work is needed to make that happen.
Early stage investments and distribution of costs—
transactional costs, fees, licensing, onboarding participants—must be accounted for. You must also understand the revenue generation component and how
members derive value from the ecosystem, balanced
with a fee model that makes the system self-sustaining.
You could implement a monthly subscription fee, a
flat monthly or yearly fee, transactional based on
volume, or other approach.

Defining your intellectual property (IP) and Data Rights Model is vital to establish early on.

Defining your intellec-
tual property (IP) and Data
Rights Model is vital to
establish early on. In today's
world, IP is a fundamental consideration for organizations to protect and/or monetize. Depending on how
you handle IP, it will either attract or detour participants from your ecosystem. You will quickly want to
answer:

- Who owns the intellectual property for the
  ecosystem?

- Who owns the data?

- How is intellectual property enforced?

Enforcement and termination focuses on the bounds within which participants can operate. The ecosystem's key decision makers must think through management of enforcement, arbitration, and termination. How will issues be identified? What are the escalation paths to decision makers? When issues arise, do you put users or companies on probation or kick them out? What happens to data and intellectual property?

There's no getting around it: selecting a business model is a tough decision. In addition to the macro-components discussed above, there are several other key levers. You want to create a model with the ability to thrive in a dynamic digital ecosystem, so the model must interact with known or new participants and maximize value for all.

RECOMMENDATIONS

1. **Are you a market driver or follower?** You need to decide if you are a market pioneer, fast follower, niche player, or passive follower of market direction. A market pioneer or fast follower requires a business ecosystem's deliberate strategy with proactive market analysis and a higher degree of risk-taking.

2. **How transparent do you want to be in the market?** The ecosystem's spectrum of awareness level in the marketplace goes from completely open to exclusive to private. This decision affects the level of competitive advantage and amount of risk, as well as combating new market entrants. Competitive advantage is also critical; your ecosystem business design requires careful consideration of the level of openness of membership and intellectual property. Some scenarios lend easily to transparency, while others are fitting for an exclusive solution for a limited group of participants.

3. **What's your level of technical capability innovation?** The level of attractiveness of your DE offerings depends on the willingness to embrace emerging technologies such as

predictive analytics, IoT, blockchain, and AI. These technologies provide many benefits when used together and offer a level of acceleration and trust to the ecosystem.

4. **Build or join?** The DE concept is being rapidly adopted by all industries around the globe, so you need to determine whether there are existing ecosystems and if so, what the risks and rewards are of creating a new DE versus joining one that already exists.

Regardless, create a minimum viable strategy. The interesting thing about DEM is that it forces you to continuously challenge previous assumptions and decisions in a constructive way. That said, instead of trying to create the perfect DE from the beginning, create a hypothesis and test it on a continuous basis to drive toward what the ideal should look like. You will find this flexibility essential as you start to answer some very challenging questions.

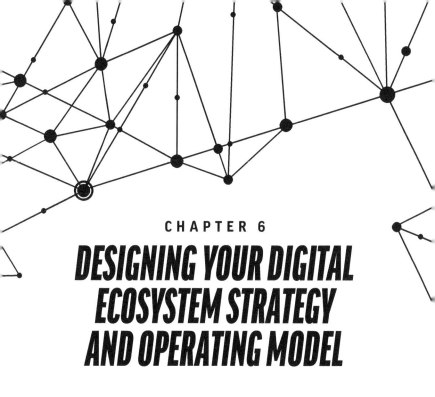

CHAPTER 6

DESIGNING YOUR DIGITAL ECOSYSTEM STRATEGY AND OPERATING MODEL

We manage a complex ecosystem across aviation and aerospace that has thousands of participants and over 21,000 independent datapoints. As we continue to digitize this traditional ecosystem to a digital one, I believe we need an open platform that creates standards while also providing the extensibility of a platform that allows our partners and even competitors to thrive within the digital ecosystem. To do this, our operating model must support a cooperative-led business model and informs the participant structure, value-exchange economics, governance, and more.

–DAVID HAVERA, BLOCKCHAIN CTO, GE AVIATION DIGITAL

A ny operating model, digital ecosystem or not, lives on essentially the same components. The second stage of DEM provides two key strategic questions to answer before proceeding to designing the ecosystem operating model: Why are we creating an ecosystem? What is the ecosystem blueprint to realize defined business outcomes? Figure 6.1 shows the digital ecosystem strategy definition leading to an operating model design—a DEM highlight reel—to identify how we realize those business outcomes in the most optimal way possible.

Some of the questions this chapter will answer include the following:

- How do we come to an operating model that makes the most sense?

- How does the strategy inform those decisions?

- How do we pay for it and keep it going?

- Who is involved in implementation and when?

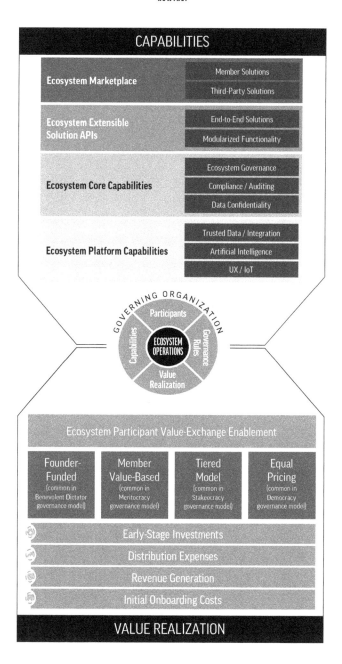

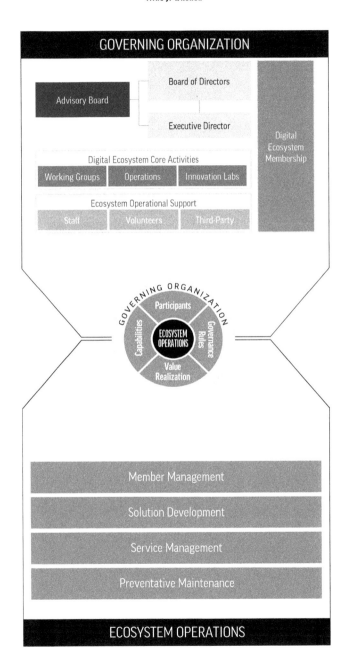

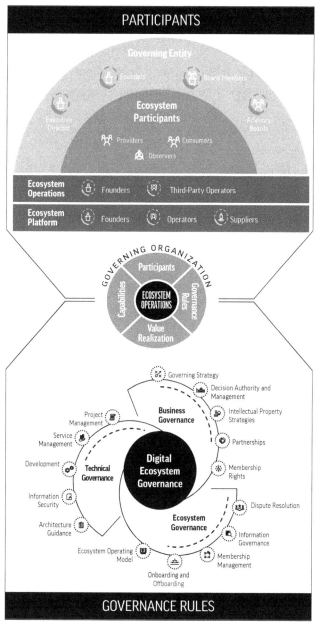

REWIRE!

Figure 6.1

65

WHAT ARE DE OPERATING MODELS?

Before we get into those questions, let's define the operating model for a digital ecosystem.

The DE operating model is like any other operating model; it is a blueprint of what will be built, derived from the DE strategy. In other words, it illustrates "how" the business model is executed through specific market and customer approaches, resourcing, technology architecture, and risk management, along with information architecture. Ecosystems can be complex, given their many different facets. That's why using the Digital Ecosystem Framework (DEF) as a guide will provide specific DE considerations to the operating model.

> **The DE operating model is like any other operating model; it is a blueprint of what will be built, derived from the DE strategy.**

The DEF provides all the vital aspects needed to define the operating model. DEF defines the governing organization, capabilities delivered, participants, value exchange, and management of the ecosystem itself. Essentially, this means it turns the strategy and business model—the "what"—into the executable how to construct the DE.

Figure 6.1 is the anchor supporting the operating

design model process and includes a few of the many options of, for example, building a governing organization for a digital ecosystem. Everything is interconnected, like a layer cake with features like strategy, business model, and governance management model. Think of it as a decoder ring for different types of organizational structures.

Here's an example: a founder-led model could be a benevolent dictator approach with a founder or group of founders in sole control, such as a business leader within organization. The most common option with a partnership-led model is two or more companies forming a steering committee that makes decisions on behalf of the ecosystem. Cooperative-led models are typically run by a board of directors that is nominated, elected, or paying a fee to be a member. Whatever the case, the group operates independently of any single organization, and the graphic shows the shift from one decision maker to a few to many.

Envisioning Insights and Digital Ecosystem Strategic Drivers

Envisioning Ideated Opportunities	Digital Ambitions	Ecosystem Business Drivers	Ecosystem Principles	Ecosystem MVE

Drives Digital Ecosystem Business Model Choice

Cooperative-Led Digital Ecosystem

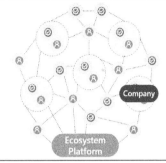

Partnership-Led Digital Ecosystem

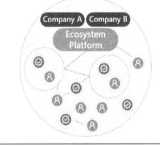

Founder-Led Digital Ecosystem

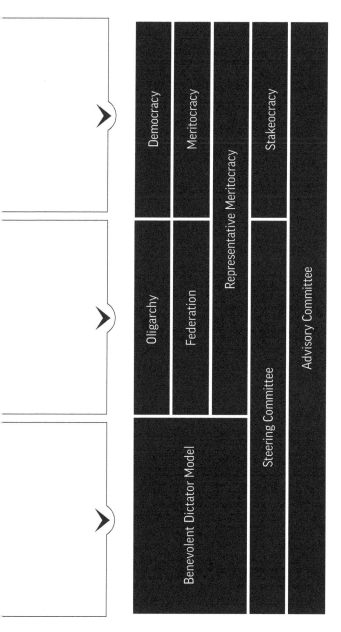

Figure 6.2

69

GOVERNING ORGANIZATION

The governing organization is the decision-making authority that defines the ecosystem strategy, governance, and policy, and provides oversight into the vision, economics, and capability development of the ecosystem.

A governing organization

- defines strategy and the economic model;

- runs decision-making committees;

- defines ecosystem capabilities;

- defines criteria for ecosystem membership;

- defines legal policies, bylaws, and regulatory compliance; and

- defines requirements for governance framework.

There are several models to choose from. Figure 6.2 is a mapping of the common governing entity management models and highlights why it is so important to start with the business model definition. It identifies where to narrow your focus, but maybe even more importantly, where not to focus.

PARTICIPATORY MODEL

The driving reason for a digital ecosystem is to create a model in which participation enables exponential value exchange through the power of connections.

Interestingly, digital ecosystem participants can play roles within multiple operational areas within the DE.

> **The driving reason for a digital ecosystem is to create a model in which participation enables exponential value exchange through the power of connections.**

Depending on the business model chosen for the DE, the level of exclusivity of your ecosystem will vary. If you choose a founder-led business model, you will have a controlled set of participants with a founder calling all the shots. However, a cooperative-led approach has the most flexibility and diversity, but governance and management will be significantly higher.

The participatory model within the ecosystem is divided into various operating areas. Governing entity operational area members have ultimate authority over the ecosystem. Common governing entities roles include:

- **Founders:** These companies invest in the establishment of the business model, defined economics, and governance of the ecosystem. Founders have the decision rights to the vision and make key decisions on what/how members are onboarded.

Participants are the core of the ecosystem and are the primary means of value exchange within the ecosystem. Common roles include:

- **Providers and consumers:** These are companies that are authorized to participate in the business benefits of the blockchain ecosystem.

- **Observers:** These companies have permissioned, read-only access to a subset of data with the ecosystem. These are often regulators, auditors, or public service data participants.

The ecosystem operations area is the accountable and responsible party for the ecosystem's preventative maintenance, service management, ongoing development, and participant management. Common roles include:

- **Third-party operator:** A company responsible for ensuring the system is performant and cost effective with the infrastructure, application platform, ledger technologies, networking, and integration operations. Operators will work with the governing entity to establish SLAs, OLAs, HA/DR, and fee structures.

ECOSYSTEM CAPABILITIES

Ecosystem capabilities are the functionality of the ecosystem. Here you will define the approach, scope, delivery, and road map within the ecosystem. In a digital ecosystem, a service-oriented-architecture approach is favorable so that the ecosystem can be robust, scalable, and as extensible as possible.

As shown in the previous graphic, these are the technical and business capability "layers" for the ecosystem:

- **Ecosystem marketplace:** Highly customizable solutions for one, many, or all participants that can be developed from the ecosystem community or third parties.

- **Ecosystem extensible solution APIs:** Composed functionality exposed either in an end-to-end functionality or as purely an application programming interface (API).

- **Ecosystem core capabilities:** Services that are required to manage, run, and extend the ecosystem.

- **Ecosystem platform capabilities:** Core technologies that enable the digital ecosystem. Specific technologies to digital ecosystems can include blockchain, AI, and IoT.

KEY CONSIDERATIONS

Regardless of how closed your business model may be, you want to design for extensibility and scalability. It is also critical to understand that technologies continually evolve, and you need the ability to adopt quickly.

Value exchange between participants will also evolve, and your digital ecosystem will need an extensibility layer open to rapid additions of functionality.

Don't target perfection right away. Address each of the four key areas in your digital ecosystem just enough to get you off the ground, creating what is known as a minimum viable ecosystem (MVE).

Embrace cloud technologies allowing you to leverage additional technology in easy and cost-effective ways. Today's technology capabilities like blockchain, AI, and IoT can rapidly accelerate the evolution of your digital ecosystem.

VALUE REALIZATION

This stage of the DEM process quantifies and defines value streams for all participants, risk tolerance levels, and digital ecosystem success criteria. In your design process, three major areas need definition:

1. **Participant value exchange:** Value exchanged by participants will vary based on the business model. You will find a combination of drivers like balancing risk versus reward and the level of cooperation you will have with potential competitors. This is quite different than what we see with other types of analog or traditional, linear business ecosystems.

2. **Ecosystem Economic Model:** This is based on intended value for participants, which is then mapped to an economic management model. Remember that it is vital to define how the ecosystem will be supported financially. We'll discuss in a moment the four primary ways in which these models can be founded that typically gravitate toward a specific business model.

3. **Expense levers:** These are the four primary levers that the economic model and value-exchange model use to define how early stage investments

will be handled, how expenses will be distributed, and what the method of revenue sharing will be.

Participant Value Exchange

The value-exchange profile looks very different in a digital ecosystem. In traditional business ecosystems, participants are grouped together in a linear, and oftentimes closed, ecosystem. Digital ecosystems are configured whereby participants can contribute, create, or consume value across any or most other participants. The notion of sequential interdependency among participants doesn't exist in a digital ecosystem. This means that participants will only be successful if the ecosystem succeeds.

Value to participants must be mapped out as a set of "gives" and "gets" across various roles. There are many tools that can help you define value for your ecosystem participants; however, an "ecosystem value map" provides a structure to outline participants' relationships, value exchanged, and information transmitted for each. You can find an ecosystem value-exchange canvas template at http://digitalecosystemplaybook.com/de-templates. Remember, when the analysis starts to benefit one organization more than others, it is time to challenge the design.

Ecosystem Economic Model

There are four primary ways in which these models can be funded:

1. **Founder-funded:** The founder manages and absorbs all costs and risks for the entire ecosystem.

2. **Member value–based:** Funded by members based on the value derived from the ecosystem. This is typically measured in the following ways: number of employees, number of transactions, or company revenue.

3. **Tiered model:** Sometimes referred to as a "pay for play" model, this funding approach establishes membership costs based on the benefits your organization wants to gain.

4. **Single pricing:** All members share expenses equally, regardless of the value they can derive, transactional volume, and the level of capability they wish to receive.

ECOSYSTEM OPERATIONS

Digital ecosystems must be continually supported after first being established. While this may be obvious, it is often overlooked or underestimated who will support the initial set of capabilities created along with the ecosystem's continuous evolution. In essence, who will operate the ecosystem and ensure a sound service management function?

> **Digital ecosystems must be continually supported after first being established.**

A clear plan is vital. Depending on the business model and other factors, it can be a variety of different roles (as defined in the participant model above).

Whether an outside party, founder, or member organization is chosen to manage the ecosystem, this party is responsible for several activities, including the following:

- Implementing and enforcing "governing organization" strategy

- Defining the operational operating procedures

- Developing and evolving ecosystem technical and business capabilities

- Creating playbooks, software development kits, and other supporting documentation

- Performing the onboarding and offboarding of participants

- Providing customer and service management

- Monitoring policy and managing compliance

MACRODESIGN MODEL RECOMMENDATIONS

1. **The business model defines the operating model:** as the complexity of models increases, transparency becomes more inclusive and time to value slows as actors change, which further cements the importance of having an informed business model conversation that dictates everything else going forward.

2. **Create a staged road map:** a digital ecosystem employs different road maps for different scenarios. Let's say you expect mass adoption very quickly. Where do investments go? How do you deal with scale and specific operators? Map your rollout to the four stages referenced earlier in the book: envision (minimum viable ecosystem), strategy and formation (production-grade ecosystem with baseline functionality), incubation (controlled rollouts of functionality), and operations (onboarding participants).

3. **Decide on an ecosystem operator early:** one particularly important issue is designating an ecosystem operator. Who will push the buttons with so many people involved? Designate someone for overall management of the ecosystem.

4. **Intellectual property:** Materially apply intellectual property and data privacy strategy to the operating model. This must be done properly to avoid increased lack of adoption. Common concerns include these: Do I own my data? Is my intellectual property protected? Who can see my data?

5. **Defined legal agreements at each stage:** For example, the envision stage should have a legal agreement (most likely nonbinding) describing your collaboration with multiple organizations to test a digital ecosystem's business viability. The strategy information period is where agreements become more concrete, with increased emphasis on protection of IP rights. The incubation stage includes finalized agreements for active participants.

Up to this point in the book, we've touched on most elements of the digital ecosystem operating model. The next chapter delves into the final component, focused around governance—the crescendo of the model that ties the business model, strategy, and operating model into one cohesive management framework.

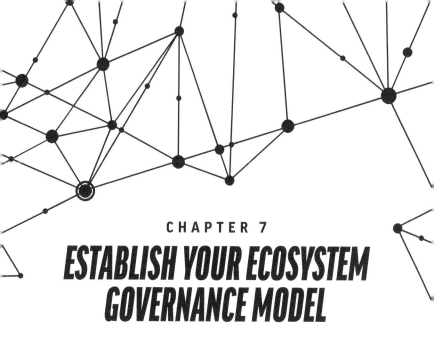

CHAPTER 7

ESTABLISH YOUR ECOSYSTEM GOVERNANCE MODEL

Through my research and partnership with top companies around the globe, I've found the most prevalent issue when establishing a digital ecosystem is navigating the complexities of all the governance permutations that are available. I've found the most effective approach is always answering what is the purpose of the ecosystem and how will it create a value-exchange network for its participants. If you are quick to jump into answering governance questions before understanding the underlying business model, you're in for a lot of complication and possibly even loss of participation in your ecosystem.

–DR. MARY C. LACITY, PROFESSOR OF INFORMATION SYSTEMS AND DIRECTOR OF THE BLOCKCHAIN CENTER OF EXCELLENCE, SAM M. WALTON COLLEGE OF BUSINESS

As Dr. Lacity explains, governance is complex and best approached by identifying ecosystem purpose. In the simplest of terms, business governance is about establishing decision authority. In a founder-led approach, for example, authority is executed by key stakeholders of the ecosystem's founding company. A partnership-led ecosystem bases authority on nominated individuals from two or more companies that form a steering committee. Cooperative-led ecosystems employ authority through traditional boardlike structures.

> **An ecosystem's governance is key when considering its capacity to execute on its purpose.**

An ecosystem's governance is key when considering its capacity to execute on its purpose. Considerations around who can become members, what rights those members enjoy, and how those members make decisions have an obvious impact on the capacity of the ecosystem to achieve its goals. These considerations drive the operations management of the ecosystem, which determines how the ecosystem handles day-to-day business and governs its approaches to both protecting and creating a thriving value-exchange network for its governing entity, participants, employees, and partners.

Onc of the largest dependencies on the governance model is the legal structure adopted by an ecosystem. This forces ecosystems down specific paths of governance, from simple guidance to very specific requirements around bylaws and how the ecosystem defines its roles and responsibilities. Often, these legal requirements are core decision points for the creators of an ecosystem, as they will include how the ecosystem deals with establishment, governing structures, participant rights, data privacy, intellectual property, and the accountability of persons or organizations.

While there are many options here, don't fret; your DE business model will inform this governance management model structure and narrow the choice from close to a dozen to a few options—or one.

Every ecosystem must be mindful of general regulatory compliance considerations, such as privacy rules and antitrust considerations. Further, an ecosystem will face additional issues to the extent of its focus on a particular industry, sector, or problem. Keep in mind that key inputs into business governance are the decisions made in the ecosystem's strategy stage—mission and vision, intellectual property, partnerships, and rights.

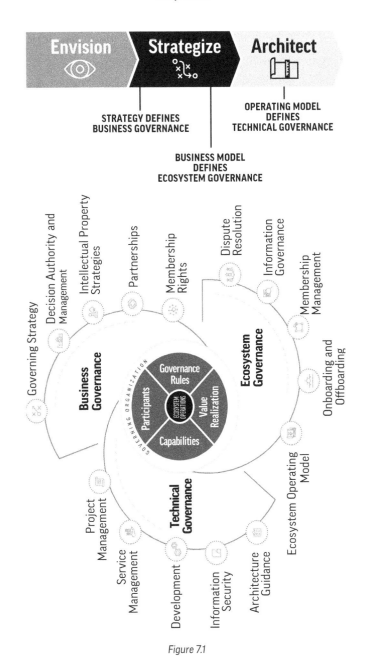

Figure 7.1

In figure 7.1 you will find the three facets of governance—business, ecosystem, and technical—that provide an understanding of the different components, when they will be addressed, resourcing participation, economics, decision-making processes, and structural configuration within the ecosystem. All three of these facets have different perspectives based on those factors.

BUSINESS GOVERNANCE

Business governance covers the ecosystem strategy, legal and compliance concerns, major decisions, and the financial model, along with the overall organizational structure. One major factor here is the governing entity management model that carries out these business governance tasks.

As defined in the last chapter, the governing organization management model structure is a common tool used by ecosystems to determine how certain actions will be undertaken or resolved among the members. This structure may be concentrated or widely disbursed among the membership, both generally and for specific types of decisions. In certain circumstances, governing boards can remove members from the board or from the consortium entirely. The ecosystem business model has defined the criteria, and the operating model has defined the actual model.

The most common activities for business governance include the following:

- **Governing strategy:** Defining the strategic element for the ecosystem by having a business model execution plan, well-formulated mission and vision statements, a plan of how the economics will work, key organizational structures, a risk management plan, and a business capability road map.

- **Decision authority and management:** Defining the leadership structure and individuals who determine how members of an ecosystem participate in formal decision-making. Forums like working groups, innovation labs, review committees, and board-of-director sessions are common.

- **Intellectual property (IP) strategies:** Creating an IP management plan identifying provisions for both the ecosystem itself and its members. This should cover how to develop, manage, protect, and potentially how to monetize IP created as a result of the ecosystem.

- **Partnerships:** Creating policies on the ecosystem's

usage of third parties, legal agreements, and the overall definition of how contractual arrangements govern the ecosystem.

- **Membership rights:** The ecosystem policies and technologies for ensuring that the proper participants have access to the right resources. This can include termination policies, arbitration, force majeure, and indemnification rights.

Business Governance Considerations

1. **Define the ecosystem critical path drivers for success.** Clearly define the business outcomes that the ecosystem will create. Ensure that these statements are quantifiable to specific metrics, qualifiable to intangible goals, are time bound, and focused on value exchange of the participants.

2. **Do your research on the jurisdiction of formation laws.** Typically, different jurisdictions have differences in their laws governing similar legal concepts (e.g., IP laws in the United States are different from those in France, although there are treaties that provide some uniformity and commonality).

3. **Some jurisdictions are cheaper than others.** Certain jurisdictions may have rules that

impose substantial costs on ecosystems or their members.

4.   **Proactively engage your legal, compliance, and risk decision makers.** Ecosystem membership should engage legal counsel regarding regulatory compliance for ecosystem operation and should be prepared to adopt compliance procedures recommended by counsel.

5.   **Your level of inclusiveness has impacts.** Ecosystems operating in concentrated industries with products available only to members may face antitrust/competition risk in the future. However, today the products/solutions offered by ecosystems are generally too undeveloped to constitute a product that is essential to effectively compete within an industry.

6.   **Don't get fancy—keep agreements clear and straightforward.** The use of simple confidentiality agreements is often considered less intimidating to signatories but may provide more rights to the party disclosing the information.

7.   **It's ideal to give members voting rights.** Voting rights are essential in determining how the membership influences the direction of the ecosystem.

8. **Don't get too fancy with voting.** An ecosystem with permissive voting rights risks inefficiencies and factionalism, while those with restrictive voting rights risk tyranny and loss of legitimacy.

9. **IP ownership should support the short- and long-term goals of the ecosystem.** Depending on these goals, patent ownership strategy may make sense as a prophylactic measure against future competitors or patent trolls and protection for the ecosystem. These protections should extend to trademarks and other nontechnical IP.

10. **Honor IP contributions.** Clear inbound and outbound contribution rights make development smoother and lower legal risk.

ECOSYSTEM GOVERNANCE

Business governance is the ultimate authority with regard to the direction and key decisions that need to be made. Ecosystem governance covers the day-to-day operations of the ecosystem. This ranges from making sure that members are happy with the ecosystem services,

> Business governance is the ultimate authority with regard to the direction and key decisions that need to be made.

the end-to-end solution is kept evergreen, disputes and arbitration are handled when issues comes up, the change management on the platform is handled, members are on- and offboarded, and the ecosystem services and solution road map are in place.

Common ecosystem governance activities include the following:

- **Ecosystem Operating Model:** Defining the overall operating roles within the ecosystem, as well as separation of concerns along with potential multiconsortia issues.

- **Onboarding and offboarding:** Bringing on new participants to the ecosystem along with removal. Should have standardized approach to business, legal, data, and infrastructure requirements.

- **Membership management:** An ecosystem must decide the types of membership it is willing to accept.

- **Information governance:** The ecosystem data protection, compliance, electronic discovery, risk management, privacy, data storage, and archiving.

- **Enforcement, dispute resolution, and litigation:** Clear definitions and rules for dispute resolution management.

Ecosystem Governance Considerations

1. **Create an ecosystem constitution agreement.** This sets the overall governance framework for the blockchain ecosystem. It addresses and settles all issues that might hamper the smooth and seamless cooperation of different members. This constitution is meant to be more than a statement of lofty aspirations; it defines the economic benefits, business considerations, governance model, and legal ramifications of membership. While a constitution is, in legal terms, a private agreement, it does provide vital decision-making information to willing participants before legal agreements are presented.

2. **Consider enlisting help from an experienced third-party ecosystem operator.** While digital ecosystems are a newer concept, traditional ecosystems run by managed service providers (MSPs) understand the governance complexities for managing multitenant-style ecosystems. This approach saves time and ultimately a great deal of headache if enlisting their help for advice or for operating the ecosystem end-to-end.

3. **Ensure that IP and data protection agreements are in place along with the technical**

controls. Classify all data that flows through the ecosystem, and map those to the proper procedures.

4. **Define the role and importance of each participant when evaluating termination rights.** Evaluate appropriate termination triggers that should be afforded to participants and that the ecosystem can exercise to remove the member. Without clear rules and parameters for termination, termination can lead to litigation, impact technological development, affect a consortium's liabilities toward third parties, and otherwise create costly negative effects.

5. **Implement continuity measures to mitigate the risk of force majeure events.** Force majeure is a standard boilerplate clause that should be scrutinized in ecosystem agreements between members and third parties. Certain critical obligations may necessitate being carved out of the force majeure provision. This can be especially true in scenarios like recent global pandemic events where, through no fault of a participant, an entire supply chain is shut down.

6. **Model the ecosystem governance based on industry dynamics.** An ecosystem should

tailor its membership interactions, structure, and working groups based upon the industry it serves and the types of participants best able to accomplish ecosystem goals and its business objectives.

7. **Careful considerations must be made for data protection, privacy, and sovereignty.** Create principles that govern how data will be handled through defined participant roles. These roles should map to specific privacy policies for each member of the ecosystem.

8. **Understand the current and future country jurisdiction legal and regulatory landscape requirements.** Aspects like audit rights and recordkeeping requirements are imposed by virtue of the ecosystem's activities and geographical reach, the composition of its members, and other factors. Conduct due diligence into industry best or suggested practices to adopt the same for compliance-driven reasons, for market perception and confidence reasons, or to mitigate future risks.

9. **Create a forum that address disputes among participants.** This process protects participants and the ecosystem as a whole from bad actors.

Providing a governance forum that separates misinterpretations or mistakes from deliberate malicious behavior is a must. Ensure that the decision makers (e.g., judges), investigators, and processes are clearly defined.

10. **Define the termination and offboarding process.** Define this process early, and ensure it is in participant agreements from the beginning. Identify whether there are any laws or regulations that if changed, repealed, or implemented would warrant dissolution or wind-down. Other factors include misusing ecosystem resources, breaking defined policies, violating data or IP policies, exhibiting poor economic performance milestones, or the participant experiencing a bankruptcy type of event.

TECHNICAL GOVERNANCE

Technical governance covers the underlying technical platform components, such as networking, infrastructure, encryption, secured environments, node management, data platforms, integration tools, and UX. Generally speaking, in a cloud governance-oriented world, many technical governance activities are consistent with those already in motion:

- **Project management:** The planning and executing of ongoing project work to meet the goals of the ecosystem. This can include resource management and project issue resolution.

- **Service management:** The aspect that designs, plans, delivers, operates, and controls the daily upkeep of the ecosystem. This can include incident management, environment monitoring, and billing.

- **Development:** Responsible for programming, documenting, testing, and bug fixing involved in creating and maintaining the ecosystem.

- **Information security:** Critical component of the ecosystem that prevents unauthorized access, use, modification, or destruction of information. This would also address governance as code.

- **Architecture guidance:** Standards, reference architecture, onboarding playbooks, and solution accelerators.

KEY CONSIDERATIONS

There are many considerations involved in the governance journey, with one consistent element: you must define the "why" of building this ecosystem, and this is articulated through a strategy of identifying an employable business model and then thinking through operational components.

The risk of redoing work significantly decreases along this path because many governance questions are answered during the process of completing strategy tasks. And you're doing it in a way that leads with a business conversation instead of a generic, out-of-context, technical, or governance conversation.

CHAPTER 8

WRAP-UP

Take a deep breath. We covered a lot of ground in this book, examining the three major facets of ecosystem governance. In the end, we have three key takeaways:

1. Next-generation business models will have digital ecosystems playing a predominant role. Enabled by emerging digital technologies, DEs will provide organizations with greater scalability, multidimensional business models, next-generation user experiences, and value exchanges that are both and resilient and secure.

2. Digital ecosystems are complex with many moving parts. Using the Digital Ecosystem

Framework, you will understand those parts and how they interrelate. This is the foundation to building your own digital ecosystems.

3. The Digital Ecosystem Method articulates the step-by-step methodology for creating a digital ecosystem in a way that is not overengineered nor underengineered.

Where do we go from here? The first step is take the DEM approach and make it your own. Reference figure 8.1 (included here with example tools) and then customize it for your needs. Add tools and techniques, add methods, remove methods. Orchestrate! Maybe you need to skip ideation altogether and go straight to strategizing. Maybe you need ideation; start there and follow end-to-end methodology. Whatever the case, make it your own mental framework to address critical questions for building a durable and robust ecosystem.

There is a lot to understand, and much of it is in flux, so start ideating now.

This is complicated. There is a lot to understand, and much of it is in flux, so start ideating now. Get your hands dirty, and even if it's just with a few people, start modeling what this looks like—get some ideas on a page, determine what they mean to you, explore the current digital ecosystem marketplace, and replicate what's out there for your business or industry.

When you're ready to dive in, take a stage-based approach. Don't try to eat the entire apple; take one bite at a time. Reference the sample implementation plan below for inspiration. Remember, the goal here is to fail fast and often, thus removing all baggage from the process immediately to know what's going to work for your ecosystem.

Like their natural counterparts outside your back door, digital ecosystems are a fascinating array of interconnected parts working in harmony. The future of business is here, and digital ecosystems are driving it. Now is the time to embrace the technological moment and lock in your company's longtime success.

Envision

PURPOSE: Ideate, envision, and identify digital ecosystem opportunities through open innovation approaches.
ACTIVITIES:

1.1 Understand both the internal needs but also the external trends in the market and the unique capabilities digital ecosystems provide.

1.3. Ideate and envision the art of the possible to identify new and unknown opportunities or refine an existing business process to maximize value.

Strategize

PURPOSE: Strategize defines the value stream that will future-proof the ecosystem business model and all major decision points.
ACTIVITIES:

2.1 Rationalize your existing business and envision the digital business strategy and ambitions.

2.2 Model the ecosystem business strategy and create the business strategy along with all the economic factors for participants.

2.3 Assess inhibitors and risks to participants and to adoption.

Architect

PURPOSE: Tuned operational model and governance with clearly defined business, technology, and human processes in place to maximize value and growth.
ACTIVITIES:

3.1 Establish the ecosystem operating model and ecosystem governance rules.

3.2 Architect ecosystem solution architecture based on the ecosystem design.

3.3 Plan the transformation road map for controlled capability rollouts to participants.

3.4 Finalize the business, ecosystem, and technical governance models.

EXAMPLE TOOLS:

Trends Radar
Business Capability Model
Journey mapping
Persona
Business scenario analysis
Strategic Alignment Matrix
Mutual non-disclosure agreement

Digital Ecosystem Training
Ideation Workshop
Disruptions Impact Analysis
Opportunities and Threats
MVE Profiling Sheet
Strategic Relevance Model

EXAMPLE TOOLS:

Ecosystem Constitution
Business Capability Model
Economic Impact Assessment
Opportunity Analysis
Opportunity / Synergy Matrix
Ecosystem Participation Matrix

Opportunity / Risk Profiling
Strategic Risk Assessments
Business Value Matrix
Risk and Information Impact Analysis
Social and Ethical Impact Road Map
Business Criticality / Realization Matrix

EXAMPLE TOOLS:

Ecosystem Architecture Document
Opportunity Trade-Off Analysis
Risk and Security Review
Option-based Road Maps
Innovative Technology Approach Matrix
Business Continuity Plan
Ecosystem Governance Model

Solution Architecture
Ecosystem Onboarding Playbook
Ecosystem Agreement / Contract
Ecosystem Reference Architecture
Ecosystem Governance Model
Consensus Practice Design

1.2 Continuous MVE   Defined MVE(s)   Iterative MVE(s) Development   Iterative MVE(s) Validation   Iterative MVE(s) Graduation

Figure 8.1

A roadmap diagram with three phases — Envisioning, Strategy and Formation, Architect and Incubate — organized by Week # columns, and two top categories COLLABORATION and STRATEGY.

Envisioning
- Founders Collaboration Agreement
- Business Strategy Alignment
- Business Envisioning
- Digital Transformation Scoping

Strategy and Formation
- Define and Establish Advisory Board (Formation Participants Collaboration Agreement)
- Recruit Advisors
- Participant MOU Agreement Created
- Ecosystem Operations Agreement Definition
- Business Model Design
- Economic Value Model
- Risk Assessment Analysis
- Change Management Planning
- Technical Capability Assessment
- Ecosystem Design

Architect and Incubate
- Participant MOU
- Recruit Early Adopters
- Participant MOU
- Recruit Early Adopters
- Ecosystem Solution Architecture
- Transformation Roadmap
- Finalize Governance Model

COLLABORATION

STRATEGY

DELIVERY

Rapidly Define and Legitimize MVE	Iterative MVE(s) Development and Validation			Productionalize MVE(s) and Graduate
Founder Steering Authority	Advisor Validation and Feedback			Change Management Established
Pilot Use Cases	Development Sprints			Controlled Rollouts of Incubation Capabilities
	Data Architecture and Policy	Security, Compliance, and Legal	Data Integration, Processing, and Ingestion Use Cases	Production Data Loads

Figure 8.2

103

RESOURCES

Interested in learning more about digital ecosystems and working with Mike J. Walker? Access these free resources online:

Subscribe to the *Digital Ecosystems Newsletter:*
http://digitalecosystemplaybook.com/subscribe

Assess your Digital Ecosystems Readiness with the Digital Ecosystems Assessment Tool:
http://digitalecosystemplaybook.com/de-assessment

Starter introduction videos:
http://digitalecosystemplaybook.com/de-videos

Starter templates:
http://digitalecosystemplaybook.com/de-templates

Get in touch with Mike:
Website: www.mikejwalk.com/contact
Twitter: @MikeJWalker
LinkedIn: https://www.linkedin.com/in/mikejwalker/

CPSIA information can be obtained
at www.ICGtesting.com
Printed in the USA
BVHW090834151220
595781BV00012B/219